Rocks & Minerals
of the Northern Rockies

A GUIDE TO GEMS, MINERALS, AND CRYSTALS

- **Quick Identification and Unusual Characteristics**
- **Appearance and Crystal Shape**
- **Color and Hardness**

David Trexler

WATERPROOF

Sedimentary Rocks

CLASTIC ROCKS

- One of the most prevalent rocks earth; can be any color, but brow common; layered in appearance
- Names based on individual gra from large to small: Cobbleston Pebblestone, Sandstone, Siltstone, and Claystone (shale); chemical composition doesn't matter for names; **Mudstone** is the term for rock containing mixed size small grains
- **Sandstone** is composed of grains 0.07–2 mm in diameter; Quartz is the most common grain type; individual grains must be cemented together, not melted together; often formed from dunes (wind) or sandbars (water)
- **Shales** are microscopic-grained Sedimentary rocks; can be from any source; identified by grain size and layering; the most common deep-ocean deposit

CARBONATE ROCKS

- Identified by chemical composition
- Usually white to gray in color
- Formed primarily in shallow to mid-depth ocean deposits (< 10,000 ft), although may be formed in fresh water as wells
- **Travertine** is a carbonate rock formed by hot spring deposits
- **Limestone** is the most common of the Carbonate rocks; bubbles or fizzes when in contact with acid; many massive limestones are actually fossil coral reef deposits; used to make the cement for concrete; a main supplement in chicken feed for egg production
- By strict definition, Calcium Carbonate, with other minerals, collectively are called carbonate rocks; often contain, or are composed of, fossils

What are Rocks and Minerals?

Minerals are solid molecular compounds. These naturally-occurring compounds form crystals if given the chance. The most common mineral in the earth's crust is silicon dioxide (SiO_2). When originally formed, it became the base compound that "floated" to the top of the molten mass of the earth, and it formed the bulk of the earliest solid crust of the planet. When silicon dioxide is allowed to cool slowly, beautiful Quartz crystals may be formed.

Rocks are solid chunks composed of one or more minerals. Thus, minerals are always discrete chemical compounds while rocks may be composed of many minerals. Commonly, rocks are what we see in road cuts and hillside exposures.

The Three Distinct Classes of Rock

SEDIMENTARY

Sedimentary rocks are those formed from consolidation of dust and debris on the earth's surface. These rocks usually form in layers and often contain fossils.

Most sedimentary rocks were laid down as marine sediment layers, so uniform, parallel layering is very common. Sedimentary rocks tend to be the softest of the three types.

METAMORPHIC

Metamorphic rocks are those which were originally formed as one of the other rock types, but which were subsequently altered and distorted through heat and pressure.

Heat-related color changes and originally round grains that have been elongated, and/or partially melted, and are some of the identifying characteristics of these rocks.

IGNEOUS

Igneous rocks are those cooled from the molten mantle of the earth. Their appearance depends on the rate of cooling and their chemical composition.

Igneous rocks are most easily identified by their angular and often crystalline structures and the lack of organic traces within the rock. They also tend to be harder than sedimentary rocks.

Rock Identification

Rocks are identified in different ways depending on type and structure. For instance, the diameter of the original grain is used to identify some sedimentary rocks. From large to small, these are cobblestone, sandstone, siltstone, and shale (also called claystone). Sandstone is formed from grains ranging in size from 1 mm down to the thickness of a human hair. The individual grains may be composed of any mineral.

Other sedimentary rocks and most igneous rocks are identified on the basis of chemical or mineral composition. Accurate identification by these characteristics almost always requires laboratory testing.

Metamorphic Rocks

GNEISS

- Opaque with some slightly translucent crystal grains
- Mixture of distorted mineral grains and crystals, often banded; hardness: 5–7, depending on composition; higher temperature metamorphic than schist, with more crystals
- Common worldwide; most common type of oldest rocks on earth; formed at crustal depths of 6–15 miles

SLATE

- Many colors depending on mineral composition; opaque
- Partially re-melted shale; hardness: 5.5
- Common worldwide; noted for its ability to separate into uniform sheets; common building stone, often used for roofing tiles

SCHIST

- Composed of flattened visible mineral grains from any original igneous or sedimentary source; hardness: ~4 to 5, varies with mineral composition
- Opaque, often shiny, flaky; often identified by its "sparkly" appearance; high grain distortion, low amount of crystal regrowth
- Common worldwide

MARBLE

- Re-crystallized limestone or dolomite; hardness: 3; color: white, but impurities produce broad color range; slightly translucent to opaque
- Common worldwide; commonly used building and statuary stone; trade name "marble" may also refer to limestone, travertine, or other stone products

QUARTZITE

- Many colors, but commonly translucent gray-brown
- Partially re-melted sandstone; hardness: ~7
- Common worldwide; commonly used for countertops

Igneous Rocks

GRANITE

- Light colored, commonly red (Iron staining in orthoclase feldspar) and white (quartz)
- Often composed of many minerals, but must mainly contain quartz and orthoclase
- Slow cooled intrusive volcanic; visible grains
- Most common igneous rock in the continental crust; often used for kitchen countertops

GABBRO

- Dark-colored, commonly greenish (olivine) and black (various pyroxenes and plagioclase feldspar), primarily composed of plagioclase feldspar and pyroxenes
- Slow cooled intrusive volcanic with visible grains; most common igneous rock in the deep ocean crust
- Often used for grave marker headstones

PEGMATITES

- The name pegmatite refers to a coarsely crystalline igneous rock with crystals an inch or more in length
- Phenocryst is the name given to an individual crystal that is significantly larger than those of the rest of the matrix
- Many types of gemstones form as phenocrysts in pegmatites and porphyries

OBSIDIAN

- Volcanic glass; usually over 70% micro-crystalline quartz
- Unstable over geologic time; glassy appearance changes to an earthy, crystalline structure
- When chipped, produces one of the sharpest cutting edges; used for surgery

BASALT

- Extrusive volcanic; chemically similar to gabbro, but rapidly cooled so any crystals formed are microscopic
- Found worldwide; more than 90% of all of earth's volcanic rock is basalt

PORPHYRY

- Similar to pegmatites, but with smaller crystals. Fine-grained volcanic rock with larger crystals up to an inch in length.
- The larger crystals are called phenocrysts regardless of composition
- Found worldwide; often used in decorative jewelry and carvings

Minerals

Mineral Identification

Chemical Composition

Chemical composition is the primary way minerals are classified. The most basic class of minerals are those made of a single element, such as gold or sulfur. Other classes of minerals include silicates, carbonates, sulfides, halides, and others. However, this guide uses the most obvious characteristics—color, hardness, and crystal shape—to provide a quick reference to the more common and interesting minerals. There are several methods of mineral classification currently in use. The most widely used is the Dana System, which contains 8 broad categories of minerals. Examples of minerals presented here are grouped within these categories.

Dana System	
1	Carbonates
2	Halides
3	Hydroxides
4	Oxides
5	Phosphates
6	Silicates
7	Sulfates
8	Sulfides

Hardness

Hardness is used as a primary way to identify different minerals. Hardness cannot be applied to rocks, since a rock is composed of multiple minerals. The Mohs' Hardness Scale is successively harder on a scale from 1 to 10 as defined by the minerals in the chart on the right.

While hardness test points are readily available, quick identification is often done by using one of the "tools" listed on the right. To test, try to scratch the material to be identified. If it scratches, the hardness is less than that of the tool used.

Mohs' Scale		
1	Talc	
2	Gypsum	
		< Fingernail (2.5)
3	Calcite	
		< Copper Penny (3.5)
4	Fluorite	
5	Apatite	
		< Knife/Glass (5.5)
6	Orthoclase	
		< Steel Nail/Tool (6.5)
7	Quartz	
8	Topaz	
		< Masonry Bit (8.5)
9	Corundum	
10	Diamond	

Native Element Minerals

Minerals composed of a single element

GOLD

- Yellow only; opaque; cubic crystals, but usually found as nuggets or flakes
- Hardness: 2.5; malleable–can be rolled thin enough to become semi-transparent
- Rare, but occurs worldwide

COPPER

- Red-orange, but rapidly oxidizes to blue or green; opaque
- Cubic crystals, but usually found as nuggets or veins; hardness: 3
- Relatively common worldwide; occurs in elemental form, but is much more commonly combined with other elements; most important electrical conductor worldwide

Carbonate Minerals

CALCITE

- Clear to white, may also be yellow or greenish; transparent to opaque
- Crystals pyramidal or rhombic; hardness: 3 (by definition)
- Crystalline form of limestone; clear form used in the Norden bombsight

AZURITE

- Light to dark blue
- Translucent to opaque
- Crystals pyramidal or tabular; hardness: 3.5–4
- Mined as copper ore

MALACHITE

- Light to dark green; translucent to opaque
- Crystals rare, tabular or pyramidal; hardness: 3.5–4
- Often found with Azurite

Sulfate Minerals

BARITE (BARYTE)

- Barite is the most commonly accepted form of the name, but international rules use the other; Colorless to white or yellow, rarely blue
- Crystals are tabular to massive, parallel to base; hardness: 3–3.5
- Usually formed in depositional or evaporitic environments, but has been found in meteorites; primarily used in the oil and gas drilling industry to add weight to drilling mud and prevent blowouts

SELENITE

- Clear, gray to greenish yellow brown; transparent to opaque
- Crystals usually tabular, may be pyramidal; hardness: 2 (by definition)
- Crystalline form of gypsum

Sulfide Minerals

CINNABAR

- Red to reddish gray; transparent to translucent
- Crystals pyramidal or tabular; hardness: 2–2.5
- Rare; only common source for mercury

GALENA

- Dull to shiny gray; opaque
- Crystals, cubes or octahedrons; hardness: 2.5–2.8
- Relatively common; often found in association with silver or zinc ores

PYRITE

- Yellow to gold; opaque
- Crystals, cubes or octahedrons; hardness: 6–6.5
- Common; nicknamed "fools gold" because it was an early common prospecting misidentification

Mineraloids

Substances that are neither rock nor mineral

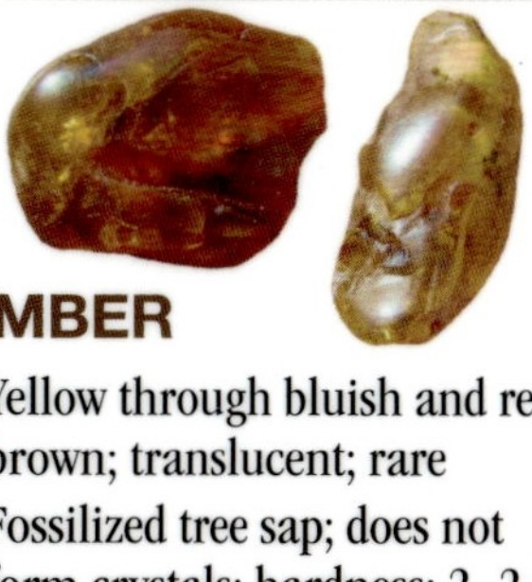

AMBER

- Yellow through bluish and red-brown; translucent; rare
- Fossilized tree sap; does not form crystals; hardness: 2–2.5

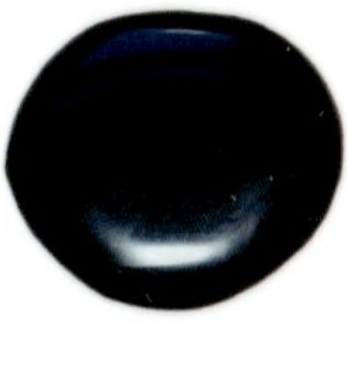

JET

- Black; opaque; common
- Gem form of lignite coal; does not form crystals; hardness: 2.5–4

QUARTZ

- Macro crystals or massive; various varieties due to impurities

VARIETY ROCK CRYSTAL

- Purest form; clear quartz or white quartz; colors caused by impurities; transparent to nearly opaque
- Hexagonal crystals; hardness: 7 (by definition)
- Relatively common worldwide

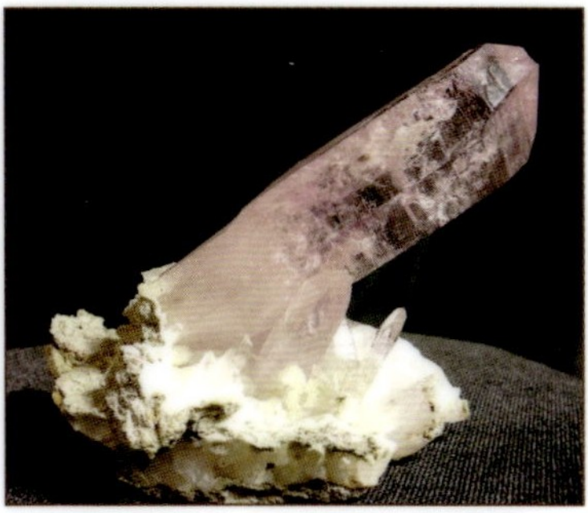

VARIETY AMETHYST

- Pinkish to violet purple; color primarily due to irradiation of iron impurities, will fade with time if exposed to UV light; transparent to nearly opaque
- Hexagonal crystals; hardness: 7
- Relatively rare, but worldwide occurrence

WYOMING JADE (NEPHRITE)

Wyoming State Mineral

- Light to dark green, nephrite worldwide can be other colors; translucent to opaque
- Crystals monoclinic and massive; hardness: 6–6.5
- Highly prized as a semiprecious gemstone

GARNET **Idaho State Gemstone**

- Worldwide, garnets can be any color, but garnets from this region are dark reddish-brown to black
- Rarely, garnet may reflect star-like light when polished (asterism)
- Usually as Dodecahedron crystals; hardness 6.5–7.5
- Idaho is one of only two places on Earth where star garnets are found. Public collection for recreational purposes at specific locations is allowed; Garnets may also be found in other localities throughout the region including along the shores of the Ruby Reservoir, Montana, and several Wyoming localities

Silicate Minerals

Most common: ~90% of all minerals in the earth's crust

CHALCEDONY

- Micro-crystal quartz with impurities
- Found in many varieties
- Most common semi-precious gemstones are found in this grouping
- NOTE: Agate and Jasper are both chalcedony. The difference is only that Agate is translucent and Jasper is opaque
- Opal is also often considered a form of chalcedony, but it is not technically a mineral because it is non-crystalline

VARIETY AGATE

One of two Montana State Gemstones

- Many colors; translucent
- Microscopic hexagonal crystals; hardness: 6.5–7
- Relatively common worldwide; primarily used in jewelry
- Prized for its unique tree-like and scenic patterns resembling those found elsewhere in nature

VARIETY OPAL

- Translucent to opaque; usually white with rainbow "fire;" lacks crystal structure—molecules form microscopic globules
- May or may not be considered chalcedony due to its lack of crystalline structure, but is chemically identical; hardness: 5.5–6
- High quality fire opal is mined near Spencer, Idaho; relatively rare worldwide

VARIETY JASPER

- Many colors; opaque
- Microscopic hexagonal crystals; hardness: 6.5–7
- Relatively common worldwide; other types of rock commonly and incorrectly called Jasper

EPIDOTE

- Green to black; transparent to nearly opaque
- Formed secondarily through metamorphism; hardness: 6–7; elongate flattened crystals, usually small
- Relatively common worldwide

TOURMALINE

- Commonly black; can be virtually any color; translucent to opaque
- Elongate, striated crystals; hardness: 7–7.5
- Relatively common worldwide; often seen as black lines in granite

Oxide Minerals

SPINEL

- Most color; transparent to opaque
- Octahedral crystals or flattened triangular plates; hardness: 7.5–8
- Relatively common worldwide; synthetic spinel common–one synthetic form is used as a replacement for window glass

CORUNDUM variety Sapphire

One of two Montana State Gemstones

- Relatively common worldwide, but gem quality stones are rare
- Occurs in most colors, transparent to opaque; red color is called Ruby instead of sapphire
- Hexagonal, columnar crystals, often flattened; hardness: 9 (by definition)
- Sapphires of varying colors may be found near Helena and Phillipsburg, Montana and public access is available; blue "Yogo" sapphires are only from one area near Hobson, Montana, and no public access is allowed

Halide Minerals

HALITE

- White in pure form, but often colored by inclusion of other trace minerals; translucent
- Cube-shaped crystals; hardness: 2.5
- Common worldwide, especially dissolved in seawater
- Mined as table salt, road salt
- Halite is listed here, even though it is not found in outcrop in the region. However, it is the primary mineral in the rock containing the huge oil and gas reserves of the Williston Basin; also occurs in mines in East-Central Idaho and Southwest Wyoming

FLUORITE

- Colorless, but often colored by inclusion of other trace minerals; translucent
- Cube-shaped crystals; hardness: 4 (by definition)
- Relatively common worldwide; has been called the most colorful mineral in the world

Rock Collecting and Identification

One of the earliest human hobbies was collecting interesting and unusual rocks and minerals. Understanding the composition, origin, and properties of rocks and minerals is vital to this process of mining—and mining produces the base materials we need for our current industrial lifestyle.

Whether you're a professional geologist looking for the next outcrop of minerals needed by industry or just a curious amateur, rocks and minerals tell an important story and interpreting that story is a necessary part of that process. While by no means comprehensive, the information in this field guide is designed to provide you with the basics to help you understand the story represented by the rocks and minerals included here.

Legal Regulations

It is generally assumed that the owner of the land owns any economically viable mineral that occurs below the surface. Gravel, limestone, and even water have been argued in court to be minerals, and there is some gray area. For most instances, if you have permission from the landowner, you may legally collect surface rocks.

Be kind to your landowner—and the land itself! Rock collectors are seeing more and more restrictions, many of which are meant to prevent littering, vehicle ruts, and wildlife disturbances. A good rule of thumb is to always leave an area in as good or better condition than it was before.

Each government agency has its own rules, depending on the land classification. It is always illegal to collect in national parks, monuments, and wilderness areas without a permit. Check with the appropriate agency or landowner for their collection rules and regulations to ensure you collect in accordance with the law.

"A rock is just a rock" unless it is something special, and you know where it came from and can return to the exact location to gather more. Our history is full of stories of discoveries made, then lost. Be sure to document your discoveries in such a manner as to be able to return to the site!

Be safe! Dangerous wildlife including spiders, snakes, bears, and mountain lions could all turn a fun rockhounding weekend into a dangerous and even deadly ordeal for someone unprepared. Changing weather, flash floods, and vehicle failures have all cost rock hunter lives.

Tips and Tricks

Get to know the area and its geology before you go. It is much more fun in the field if you know what has been found in the area. Area geology, color, hardness, and shape will allow most common rocks and minerals to be identified in the field.

Pack adequate gear. For most rockhounding expeditions, an empty backpack going out translates into a full one coming back! Tools like rock hammers, GPS units, notebooks, water and energy foods should always be a part of any field kit.

Familiarize yourself with rock and mineral identification. Unfortunately many rock types grade into each other, so precise identification is often left until lab equipment is available.

Have Fun!

MADE IN THE USA
MANAGED FOREST
VEGETABLE OIL INK
$8.95
ISBN 978-1-954018-08-2
50895>
www.QuickReferencePublishing.com

(561) 733-2883 #355 0326
9 781954 018082